A Critical Examination of Classical and Quantum Mechanical Waves

Greg Feild

June 18, 2017

ripple:

    ripple
    in still water
    where there is no
    pebble dropped,
    nor wind to blow

          -- Robert Hunter,
           (The Grateful Dead)

Abstract:

In this paper, we demonstrate there are no classical or quantum mechanical waves of any description.

In our universal model of the sinister universe, there is no waving, undulating, or stretching; of space, or time, or matter.

About the author:

Greg Feild enjoys physics as his vocation,
and philosophy as an avocation.

Philosophy, as we use the word, is a fight against the fascination
which forms of expression exert on us.

-- Ludwig Wittgenstein

Introduction:

Our new model of the world is a mechanical model; albeit without the gears, wheels, and pulleys, that Maxwell and Faraday imagined.

Fundamental particles are spinning mass-energy. Half integral spin is the origin of inertial mass. The mass-energy of a particle is 'composed' entirely of rotational kinetic energy. This rotational kinetic energy is due to the particle's spin, *and* spin precession about the direction of the linear momentum of the particle.

The linear momentum of a particle (or more specifically, impulse) is due to a particle's helicity flipping harmonically at the characteristic frequency that defines the energy (and 'wavelength') of the particle.

This is true for the photon, as we have argued previously, as well as for the leptons, as we shall demonstrate in this paper.

All particle interactions are ultimately characterized by, and completely determined by, the exchange of energy, momentum, and *angular* momentum; conserving all three, of course!

In the final analysis, all conserved quantities can be ultimately described, reduced to, or explained as a consequence of *the conservation of angular momentum*.

Particle in a box:

Let's begin with the classic example of the particle in a one dimensional box.

A particle (e.g. an electron) in a box bounces back and forth between the walls. The particle follows a well defined path and physically traverses *all points* lying between the walls of the box.

The current interpretation of quantum mechanics posits there are points where the electron never is; points the electron does not pass through; points along the trajectory where the electron essentially ceases to exist!

Of course, these points are the nodes where the quantum mechanical wave function of the particle (a sine wave) passes through zero.

In our new model, these nodes represent points where the electron is actually *physically unable* to interact because it is undergoing a 'spin flip', or a change in helicity.

We can now understand the spinor nature of the electron, and why it's spin is 'double valued'.

As the electron travels along, its spin precesses once about the direction of travel (it's 'proper' or measured helicity, say) and then the spin flips, going around once the other way; ad infinitum.

At the instant of the helicity flip, the electron is not able to interact. These points correspond to the nodes of the quantum mechanical wave function.

We can now interpret the probabilities generated from squaring the wave function to indicate where a particle is able to physically interact and *be measured*, rather than where the particle is.

In addition, like a swimmer doing laps in a pool, the boundary conditions of our box dictate that the electron must perform a helicity flip when it bounces off the walls.

This also explains the nature of the Bohr orbits in the hydrogen atom, and why they must correspond to integral values of the electron wavelength.

Way.

The wave function:

The wave function is longer mysterious or spooky.

You can no more predict the position of an electron bouncing around in a box than you can predict the position of an ideal gas molecule bouncing around in a box.

The wave function is a mathematical function invented by human beings. It has nothing to do with the particle 'itself'. The wave function is *not* the particle and does not 'collapse' when one makes a measurement anymore than a momentum vector describing the particle would collapse.

The wave function is a probabilistic determination of what energy, momentum, and position values we can expect to measure for a particle *we cannot see*.

The wave function is not a real, physical thing. It has nothing whatsoever to do with the particle; either physically, or 'metaphysically'.

The wave function does collapse.

It is a completely meaningless notion and a *non problem*.

Quantum uncertainty:

I do not believe the following thought is original to me.

Although the predictions of quantum mechanics are probabilistic; the predicted values for the measurement of a variable, or the outcome of an experiment, are *exact*.

In this sense, indeterminacy actually disappears in quantum mechanical calculations.

There is no room for the uncertainty or 'chaos' plaguing classical calculations due to the impossibility of our ever completely specifying or controlling the initial conditions of an experiment.

Photons and electromagnetic waves:

The classical electromagnetic wave is *also* the quantum mechanical wave function of the photon.

Do people worry about the collapse of the electromagnetic wave?

Classically, electromagnetic waves impinge on an electron, causing ti to oscillate back and forth.

However, "we now know" this is not the correct picture. Oscillating electrons are actually absorbing and emitting photons, and the electromagnetic wave equation describes the behavior of the photons.

The total energy of the electromagnetic wave, calculated using classical field theory, is equivalent to the energy of an integral number of photons of energy, $E = hbar*nu$, where nu is the frequency of the electromagnetic wave.

This is your correspondence and complementary principles, rolled into one.

The speed of light:

As we showed in "Reflections on the Sinister Universe ", a particle cannot exceed the speed of light because particle interactions cannot exceed the speed of light.

Interaction with other particles is what causes acceleration.

We can employ similar arguments to explain why all observers measure the same value for the speed of light.

This is left as an exercise for the reader.

Out of respect   :)

(Plus, we're growing tired of the sound of our own voice.)

(And the royal we!)

(The "we" is really meant more in the sense of we are working things out together!)

(And now we've just filled another page!)

(More or less.)

Weak isospin:

In "On Current Physics", we introduced the two fundamental leptons; the charged and neutral tetrahedron.

Geometric rotations of the three principle axes of inertia of the tetrahedron about the axis of *spin*, generate the three lepton families, transforming one into another; all manifestations of a single particle.

These rotations correspond to the transformation matrices of the group SU(3) of the standard model.

Weak isospin rotations transform the neutral tetrahedron into the charged tetrahedron.

It seems these do not correspond to any real, physical, rotation.

We speak metaphorically of the electron as an excited neutrino; its electric charge a consequence of quantum mechanical electromagnetic induction (qemf).

However, qemf is quantized, so the electron cannot decay or collapse into a neutrino.

Charged leptons can 'decay' into one another, but the charged tetrahedron is conserved.

In conclusion, there are two fundamental leptons. One is neutral and one is electrically charged.

Particle families:

We've done a lot of hand waving, and hand wringing, concerning the nature of the mass hierarchy of the three particle families.

A clear picture has yet to emerge. (Probably because it will involving doing math.)

However, it seems the solution will be that nature always conserves the 'particle lagrangian' during a decay, and hence the action. Or something.

Particle lagrangian $= m - m_0 =$ const.

The Dirac equation:

Why is the normalization for a Dirac spinor $1/(E + m)^{\wedge}\frac{1}{2}$?

This is quite unsettling since the normalization is greater than the total energy of the particle by an additional amount corresponding to the rest mass of the particle.

This cannot be right!

The Dirac equation for a free electron is

$$H \psi = (\alpha \cdot p + \beta m_0) \psi \qquad (1)$$

where the Hamiltonian H, is the total energy of the electron, and m_0 is the rest mass. The Hamiltonian operator is

$$H = i \, \partial / \partial t \qquad (2)$$

In our model, the mass-energy, $E = mc^{\wedge}2$, is the particle's coupling charge, and completely determines the particle's behavior. So the rest mass term in the Dirac equation seems, somehow, 'redundant', unnecessary, and *unwanted*.

Let's define the kinetic energy operator (T = m - m_0) to be

$$T = H - \beta m_0 \qquad (3)$$

We still don't like, or want, this extra, constant, additive, and *global* charge mucking up our beautiful equation, so (and I think you see what's coming ...) we make a *global gauge transformation* and the problem is gone!

The Dirac equation for a *massive* electron is now

$$T == H \psi = (\alpha \cdot p) \psi \qquad (4)$$

and the spinors for the left handed and right handed solutions are now *decoupled*.

If we add interactions to the mix, the general form of the Dirac equation will be

$$(T - V) \psi \;\; == L \psi = 0 \qquad (5)$$

where, of course, L is the Lagrangian operator.

Gauge invariance:

The universal model takes advantage of, and explains, both global and local gauge invariance.

Global gauge invariance allows us to remove the rest mass of a particle from our equations, because it is a global, constant, and therefore uninteresting, contribution to the particle's variable mass-energy charge.

Local gauge invariance is necessary because a particle's mass-energy charge varies during an interaction.

The eternal universe:

We predict the picture of the universe that will emerge after a reevaluation of the current cosmological data, will be one of an eternal, infinite, and *static* universe.

From lowly elementary particle decays to exploding stars, the universe is constantly being reseeded with new materials to build new stars, new solar systems, new galaxies.

Conclusion:

This will be our last book on physics for a while.

I know it is a little bit short, but we are plum out of ideas, and a little tired.

Happy, but tired!

Zero point energy:

There are no fields. There is no vacuum.

There is no zero point energy.

physics is fun!

Our dim dark age:

In the past one hundred years, physicists have set humanity back one thousand years.

Scholastic speculations on multiple universes, many worlds, the collapse of the wave function, etc., etc., are as meaningless as wondering how many angels can dance on the head of a pin.

Tragically, these ridiculous ideas have permeated modern thought and completely saturated our society. People will believe *anything*.

It will take another thousand years to undo the damage.

Let us return to the values of the Enlightenment.

The veil of the temple:

Forgive them, for they know not what they do.

(mixing metaphors like neutrinos!)

philosophy is fundamental

References:

Modern Elementary Particle Physics
Gordon Kane

Classical Dynamics of Particles and Systems
Jerry B. Marion

Foundations of Electromagnetic Theory
John R. Reitz, Frederick J. Milford, Robert W. Christy

Quantum Physics
Rolf G. Winter

Gauge Theories in Particle Physics
I. J. R. Aitchison and A. J. G. Hey

Quarks and Leptons: An Introductory Course in Modern Particle Physics
Francis Halzen, Alan D. Martin

Quantum Field Theory
F. Mandl, G. Shaw

Symmetries and Group Theory in Particle Physics
Giovanni Costa, Gianluigi Fogli

and

Elementary Modern Physics   (Best Book Ever!)
Richard T. Weidner, Robert L. Sells

a feild theory   :)

Books by Greg Feild:

<u>the pentateuch</u>

1. "A quantum mechanical theory of gravitational interactions"
   CreateSpace Independent Publishing, 8/29/2016

2. "Observations on the quantum mechanical nature of gravity"
   CreateSpace Independent Publishing, 10/8/2016

3. "On gravitation and electric charge"
   CreateSpace Independent Publishing, 11/1/2016

4. "On spin, mass, and charge"
   CreateSpace Independent Publishing, 11/29/2016

5. "On angular momentum, acceleration, and absolute motion"
   CreateSpace Independent Publishing, 1/4/2017

<u>the exegeses</u>

6. "The Sinister Universe"
   CreateSpace Independent Publishing, 3/1/2017

7. "On Parity and Isospin"
   CreateSpace Independent Publishing, 4/11/2017

8. "Reflections on the Sinister Universe"
   CreateSpace Independent Publishing, 5/12/2017

<u>the hermeneutics</u>

9. "On Current Physics"
   CreateSpace Independent Publishing, 6/11/2017

do physics!

Notes:

a pipe dream   :)

www.ingramcontent.com/pod-product-compliance
Lightning Source LLC
Chambersburg PA
CBHW072049190526
45165CB00019B/2247